NATIONAL GEOGRAPHIC KiDS

美国国家地理 趣味双语小百科

原版引进
汉英对照

大熊猫和可爱的朋友们

The Panda and Lovely Friends

[美] 美国国家地理杂志儿童版 / 编写　付灿 / 译

四川少年儿童出版社

鬃狮蜥因其喉下及颈部的棘状鳞而得名。

The bearded dragon gets its name from the row of spiny scales around its throat and neck.

当当当！

KNOCK,KNOCK.

Who's there? Emma.
Emma who? Emma going to the store, do you need anything?

谁在敲门？ 埃玛。
哪个埃玛？ 我要去超市，你有什么要捎带的吗？

*Emma（埃玛）谐音I am（我是）。

ENCYCLOPEDIA OF ANIMALS 动物小百科

眼镜猴的一颗眼珠就比它的大脑大。

Each of a tarsier' s eyeballs is larger than its brain!

Q What do pigs wear on their feet when it' s cold out?

天冷的时候，猪脚上穿什么？

A Pigskin boots.

猪皮靴。

当当当！
KNOCK,KNOCK.
Who's there? Berry.
Berry who? Berry nice to see you again.
谁在敲门？ 贝里。
哪个贝里？ 很高兴再次见到你。
*Berry（贝里）谐音very（很）。

当当当！

KNOCK, KNOCK.

Who's there? **Knee.**
Knee who? **Knee to come in, it's cold out here!**

谁在敲门？ 尼。
哪个尼？ 快进去，外面好冷！

*Knee（尼）谐音need（需要）。

乌龟会成群结队地形成乌龟群。

A group of turtles is called a bale.

ENCYCLOPEDIA OF ANIMALS 动物小百科

黑猩猩和人类是近亲，它们与人类的基因相似度高达98%！

Chimpanzees are our closest relative – we share 98 percent of the same DNA!

TONGUE TWISTER 英语绕口令

Say this fast three times
快速念3遍

Are our oranges ordered?

当当当！

KNOCK,KNOCK.

Who's there? Tuna.
Tuna who? Tuna the TV, my show is starting!

谁在敲门？ 图纳。
哪个图纳？ 开一下电视，我的节目要开始了！

* Tuna（图纳）谐音turn on（打开）。

当当当！
KNOCK, KNOCK.

Who's there? Minnow.
Minnow who? Let minnow when I can come in!

谁在敲门？ 米诺。
哪个米诺？ 告诉我啥时候能进去！

*Minnow（米诺）谐音me know（我知道）。

ENCYCLOPEDIA OF ANIMALS 动物小百科

海鸥能喝盐水！这是因为它们体内有种特殊的薄膜能够过滤盐分，然后将盐分通过喙上的小口排出。

Seagulls can drink salt water! Special glands flush out the salt. It is then forced out of openings in their beaks.

Q

What do you call a dog who talks a lot?

说个不停的狗叫什么？

A

A labrador.
拉布拉多犬（Labrador“拉布拉多”谐音blabla“长篇大论的废话、碎碎念”）。

ENCYCLOPEDIA OF ANIMALS 动物小百科

老虎一餐的食肉量高达70磅（约32千克）。

A tiger can eat up to 70 pounds of meat in one meal.

Q Why do ghosts like to ride in **elevators?**

鬼为什么喜欢乘电梯？

A 因为灵魂可以飞升。

Because it lifts their spirits.

当当当！

KNOCK,KNOCK.

Who's there? Lettuce.
Lettuce who? Lettuce in,we have to finish dressing.

谁在敲门？ 莱蒂斯。
哪个莱蒂斯？ 快让我们进去，我们得把衣服穿好。

*Lettuce（莱蒂斯）谐音let us（让我们……）。

小熊猫和猫一样，通过舔舐自己来清理毛发！

Red pandas clean their fur like cats — by licking themselves!

当当当！

KNOCK,KNOCK.

Who's there? **Turnip.**
Turnip who? **Turnip the heat. It's freezing in here!**

谁在敲门？ 特尼普。
哪个特尼普？ 把温度调高一点儿，这儿都结冰了。

*Turnip（特尼普）谐音turn up（调高）。

当当当！

KNOCK,KNOCK.

Who's there? Julian.
Julian who? Just Julian, there is no one else with me.

*谁在敲门？*朱利安。
*哪个朱利安？*就是朱利安啊，就我一个人。

ENCYCLOPEDIA OF ANIMALS 动物小百科

奶牛每天的饮水量高达35加仑（约132升），食量高达50磅（约23千克）！

Cows can drink up to 35 gallons of water and eat up to 50 pounds of food every day!

Q Why can't snails play nicely together?

蜗牛为什么不能在一起玩耍?

A Because they always slug each other.

因为它们总是互相攻击，都觉得自己才是最棒的（"蜗牛"谐音"我牛"）。

ENCYCLOPEDIA OF ANIMALS 动物小百科

猫利用胡须测量洞口的宽窄，判断自己是否能够进去。

Cats use their whiskers to " feel " whether they will fit into a space they are entering.

Q What is a **duck's** hate **junk food?**

鸭子最讨厌吃什么东西？

A Duck neck.

鸭脖。

当当当！

KNOCK,KNOCK.

Who's there? **Noah.**
Noah who? **Noah good place for dinner? I'm starving!**

谁在敲门？ 诺亚。
哪个诺亚？ 你知道哪儿有好吃的饭馆吗？我快要饿死了！

*Noah（诺亚）谐音know（知道）。

ENCYCLOPEDIA OF ANIMALS 动物小百科

环尾狐猴的尾部有臭腺，它们用尾巴蹭树留下独特的气味，让其他狐猴知道自己来过这儿。

Ring-tailed lemurs have a scent gland on their tails that they rub against trees. It leaves a smelly mark to tell other lemurs where they've been!

TONGUE TWISTER 英语绕口令

Say this fast three times
快速念3遍

A real rare railroad.

当当当！
KNOCK,KNOCK.
Who's there? Macaw.
Macaw who? Macaw won't start. Can I use your phone?
谁在敲门？麦卡。
哪个麦卡？我的电话打不了，能用一下你的吗？
*Macaw（麦卡）谐音my call（我的电话）。

当当当！

KNOCK, KNOCK.

Who's there? **Hugo.**
Hugo who? **Hugo to the store and I'll stay home.**

谁在敲门？ 雨果。
哪个雨果？ 你去超市，
我待在家里。

*Hugo（雨果）谐音you go（你去）。

二趾树懒在树上生活，它们以“慢”著称，大部分时间都在睡觉。

Known for their slowness, two-toed sloths live in trees. They spend a lot of their time sleeping.

当当当！

KNOCK, KNOCK.

Who's there? Art.
Art who? Art you going to open this door?

谁在敲门？ 阿特。
哪个阿特？ 你不打算开门吗？

*Art（阿特）谐音aren't（不是）。

ENCYCLOPEDIA OF ANIMALS 动物小百科

变色龙的颜色变化取决于环境、身体状况、心情及温度等因素。

Chameleons can change color quickly depending on their health, mood, or temperature.

TONGUE TWISTER 英语绕口令

当当当！

KNOCK,KNOCK.

Who's there? **Doughnut.**
Doughnut who? **I doughnut belong here. I must have the wrong address.**

谁在敲门？ 多纳特。
哪个多纳特？ 我不该来这里，我一定是走错地方了。

*Doughnut（多纳特）谐音do not（不是）。

狮子一天的睡眠时间可达20个小时！

Lions can sleep for up to 20 hours a day!

绿蛙的寿命可达10年！

Green frogs can live for up to ten years!

当当当！

KNOCK, KNOCK.

Who's there? Santa.
Santa who? Santa card to you last week, did you get it yet?

谁在敲门？ 桑塔。
哪个桑塔？ 上星期我给你寄了贺卡，你收到了吗？

*Santa（桑塔）谐音sent（寄）。

猫鼬眼睛周围的黑色区域就像太阳镜，能够减弱射入眼内的强光。

The dark areas around a meerkat's eyes work as built-in sunglasses to reduce the sun's glare on its face.

当当当！

KNOCK, KNOCK.

Who's there? Gnome.
Gnome who? There's no place like gnome.

谁在敲门？葛罗姆。
哪个葛罗姆？没有任何地方会和家一样。

*Gnome（葛罗姆）谐音home（家）。

ENCYCLOPEDIA OF ANIMALS 动物小百科

鹈鹕利用嘴下面的喉囊捕鱼。

A pelican uses the pouch under its bill to scoop up fish.

TONGUE TWISTER 英语绕口令

Say this fast three times
快速念3遍

Big fat flat black cat.

当当当！

KNOCK,KNOCK.

Who's there? Sapphires.
Sapphires who? Sapphires started. Let's make smores!

谁在敲门？ 萨菲尔兹。
哪个萨菲尔兹？ 火烧起来了。
我们让它烧得更旺一点儿！

*Sapphires（萨菲尔兹）谐音a fire（火）。

当当当!

KNOCK, KNOCK.

Who's there? Vadar.
Vadar who? Vadar, there's a fly in my soup.

谁在敲门? 瓦达。
哪个瓦达? 服务员，我的汤里有只苍蝇。

*Vadar（瓦达）谐音waiter（服务员）。

ENCYCLOPEDIA OF ANIMALS 动物小百科

红鸢妈妈一旦觉察到危险，就会发出信号让幼鸟装死，直至危险过去。

If a mother red kite eagle senses danger, she will signal her hatchlings to play dead until the coast is clear.

Q How many jugglers does it take to screw in a light bulb?

拧紧一个灯泡，需要几个杂耍艺人？

A One, but it takes three light bulbs.

一个，但是得要三个灯泡。

眼镜凯门鳄属于体型较小的一种鳄鱼。雌性眼镜凯门鳄一次能产40枚蛋。

The spectacled caiman is a smaller species of crocodile. Female spectacled caimans lay up to 40 eggs at a time.

当当当！

KNOCK,KNOCK.

Who's there? Accordion.
Accordion who? Accordion to the weather report it is going to rain today.

谁在敲门？ 阿科丁。
哪个阿科丁？ 天气预报说今天会下雨。

*Accordion（阿科丁）谐音according to（根据……）。

ENCYCLOPEDIA OF ANIMALS 动物小百科

美洲豹皮毛上的斑点又被称为玫瑰花结。

The spots on a Jaguar's fur are called rosettes.

Q What's worse than finding a worm in your apple?

比发现苹果里面有一条虫子更糟糕的事情是什么？

A Finding half a worm!

发现里面只有半条虫子！

当当当!

KNOCK, KNOCK.

Who's there? Bunny.
Bunny who? Is some bunny going to let me in?

谁在敲门? 邦妮。
哪个邦妮? 有人要让我进去吗?

*Bunny（邦妮），some bunny 谐音somebody（有人）。

HA!
HA!
HA!

马通过嘶鸣来相互交流。

Horses neigh, or whinny, to communicate with other horses.

当当当！

KNOCK,KNOCK.

Who's there? **Howard.**
Howard who? **Howard I know?**

谁在敲门？ 霍华德。
哪个霍华德？ 我怎么知道？

*Howard（霍华德）谐音how would（怎么会……）。

ENCYCLOPEDIA OF ANIMALS 动物小百科

海豚不是通过嘴呼吸，而是通过头顶的呼吸孔呼吸。

Dolphins can't breathe through their mouths. They can only breathe through the blowholes on top of their heads.

Q What kind of dog cries all the time?
哪种狗一直在哭？

A A Chi-hua-hua.
吉娃娃（哇哇）。

当当当！

KNOCK,KNOCK.

Who's there? Water.
Water who? Water you saying? I don't understand you!

谁在敲门？ 沃特。
哪个沃特？ 你说什么？我听不懂！

*Water（沃特）谐音what are（是什么）。

六斑刺鲀在遇到危险时，会迅速吞水或空气使身体膨大，并竖起棘刺以威吓敌人。

When threatened, the long-spine porcupinefish swallows water to inflate itself. Its spines stick up and make it look scary.

当当当！

KNOCK, KNOCK.

Who's there? **Adore.**
Adore who? **Adore closed, but the window is still open.**

*谁在敲门？*阿多尔。
*哪个阿多尔？*门关上了，不过窗户还开着。

*Adore（阿多尔）谐音a door（一扇门）。

当当当！

KNOCK, KNOCK.

Who's there? Paul.
Paul who? Paul up a chair and I will tell you.

谁在敲门？ 保罗。
哪个保罗？ 拖把椅子过来，我慢慢跟你说。

*Paul（保罗）谐音pull（拉、拖）。

ENCYCLOPEDIA OF ANIMALS 动物小百科

鹦鹉喜欢吃种子、坚果和水果。一些鹦鹉的寿命长达80年甚至更久！

Parrots eat seeds, nuts, and fruit. Some parrots can live for 80 years or more!

Q Why are fireflies like cars?
为什么萤火虫像汽车？

A 因为它们都有尾灯。
Because they both have tail lights.

当当当！

KNOCK,KNOCK.

Who's there? Ant.
Ant who? Ant you glad to see me?

谁在敲门？安特。
哪个安特？见到我，你难道不开心吗？

*Ant（安特）谐音aren't（不是）。

ENCYCLOPEDIA OF ANIMALS 动物小百科

法国斗牛犬实际上起源于英国。人们让英国斗牛犬与其他品种的犬杂交，最终得到了这种小型斗牛犬。

The French bulldog is actually from England! The breed was started to create a smaller type of bulldog.

TONGUE TWISTER 英语绕口令

Say this fast three times
快速念3遍

Betty baked big batches of baked beans.

当当当！

KNOCK, KNOCK.

Who's there? **Spoon.**
Spoon who? **See you spoon!**

谁在敲门？ 斯普恩。
哪个斯普恩？ 待会儿见！

*Spoon（斯普恩）谐音soon（尽快）。

ENCYCLOPEDIA OF ANIMALS 动物小百科

野生大熊猫基本上只吃一种植物——竹子！

Wild pandas eat mostly just one plant: bamboo!

Q

Why was the math book depressed?

数学书为什么抑郁啦？

因为自身的问题太多啦。

Because it had a lot of problems.

A

当当当!

KNOCK, KNOCK.

Who's there? **Butter.**
Butter who? **Butter hurry up, we are going to be late.**

谁在敲门? 巴特。
哪个巴特? 最好抓紧点儿，不然我们就迟到了。

*Butter（巴特）谐音better（更好）。

蜜袋鼯是生活在大洋洲的一种有袋类动物。它们食水果及花蜜，前后腿之间的皮肤薄膜能帮助其在树枝间滑翔。

The sugar glider is an Australian possum that glides from tree to tree using its loose skin. It eats fruit and nectar.

当当当！

KNOCK, KNOCK.

Who's there? Sherwood.
Sherwood who? Sherwood like to come inside.

谁在敲门？ 舍伍德。
哪个舍伍德？ 她想进去。

*Sherwood（舍伍德）谐音she would（她想……）。

ENCYCLOPEDIA OF ANIMALS 动物小百科

捻角羚是生活在非洲东南部的大型羚羊。

Kudus are antelopes native to eastern and southern Africa.

Q

What do you call **a bird** with no **eye?**

没有眼睛的鸟叫什么？

Dark.

乌。

A

ENCYCLOPEDIA OF ANIMALS 动物小百科

企鹅不会飞翔，但擅长游泳。它们用有力的鳍状前肢推动自己在水中快速游动。

Penguins can't fly but they're excellent swimmers. They use their powerful flippers to propel themselves through water.

TONGUE TWISTER 英语绕口令

Say this fast three times
快速念3遍

Rivers of rainwater ran through the ramparts.

当当当！
KNOCK, KNOCK.

Who's there? **Amal.**
Amal who? **Amal ready to go!**

谁在敲门？ 阿马尔。
哪个阿马尔？ 我已经准备出发了！

*Amal（阿马尔）谐音I'm，Amal ready（阿马尔准备……）谐音I'm already（我已经准备……）。

ENCYCLOPEDIA OF ANIMALS 动物小百科

海鳗身上分泌的黏液能保护其免受寄生虫的侵扰。

The body slime of moray eels protects them against parasites.

Q

Why do ducks make good polemicists?

鸭子为什么能成为杰出的辩论家?

A

Because it' s mouth too hard.

嘴硬。

当当当！

KNOCK, KNOCK.

Who's there? Rufus.
Rufus who? Rufus leaking!
Quick get a bucket!

谁在敲门？鲁弗斯。
哪个鲁弗斯？屋顶在漏水！
快拿个桶来！

*Rufus（鲁弗斯）谐音roof's（屋顶正在……）。

猪的嗅觉特别灵敏，它们用突出的口鼻寻找地上的食物。

Pigs have an excellent sense of smell. They use their snouts to find food in the ground.

当当当！

KNOCK,KNOCK.

Who's there? **Shirley.**
Shirley who? **Shirley we have heard enough jokes by now!**

谁在敲门？ 雪莉。
哪个雪莉？ 真的，我们听的笑话已经够多了！

*Shirley（雪莉）谐音surely（无疑，真的）。

当当当！

KNOCK,KNOCK.

Who's there? Bison.
Bison who? Bison, Mommy is off to work now.

谁在敲门？ 比森。
哪个比森？ 再见，儿子。
妈妈现在要去上班啦。

*Bison（比森）谐音bye, son（再见，儿子）。

猫跳跃的高度可以达到它身长的5倍。

A cat can jump about five times its own length.